Steven School.

Aurum Solis
Gold Of The Sun.

STEVEN SCHOOL.

Steven School.

Copyright © 2012 Steven School.

All rights reserved.

ISBN: 1492842974
ISBN-13: 9781492842972

Steven School.

DEDICATION

I dedicate this book to the lovers of wisdom.

DISCLAIMER.

This book is written for informational purposes only, in the interest of preserving knowledge and history. No warranty is implied or expressed, neither the author, nor the publishers of this book assume any liability for any action or misuse of this book, or the knowledge that it contains.

Steven School.

Steven School.

CONTENTS

Aurum Solis.

Gold Of The Sun.

Steven School.

Steven School.

ACKNOWLEDGMENTS

I would like to acknowledge the ancient sages, who left enough writings for us to rekindle the flame and rediscover this once lost and ancient art.

Steven School.

1 GOLD OF THE SUN.

Proverbs 3:16

Blessed is he who finds wisdom.
She is more precious than jewels,
And nothing you desire compares with her.
Long life is in her right hand,
In her left hand are riches and honor.
Her ways are pleasant and all her paths are peace.

THE SEALS.

1. And I saw when the lamb opened one of the seals, and I heard as it were the noise of thunder, and one of the four beasts said, come and see.

2. And I saw, and behold a white horse, and he that sat on him had a bow, And a crown was given unto him, and he went forth to conquer.

3. And when the lamb had opened the second seal, I heard the second beast say, come and see.

4. And there went out another horse that was red, and power was given to him that sat thereon to take peace from the earth, and that they should kill one another, and there was given unto him a great sword.

5. And when the lamb had opened the third seal, I heard the third beast say, come and see. And I beheld, and lo, a black horse, and he that sat on him had a pair of balances in his hand.

6. And I heard a voice in the midst of the four beasts say, a measure of wheat for a penny, and three measures of barley for a penny, and see thou hurt not the oil and the wine.

7. And when the lamb had opened the fourth seal, I heard the voice of the fourth beast say, come and see.

8. And I looked, and behold a pale horse, and his name that sat on him was death, and hell followed with him, and power was given unto them over the fourth part of the earth, to kill with sword and with hunger, and with death, and with the beasts of the earth.

9. And when the lamb had opened the fifth seal, I saw under the altar the souls of them that were slain for the word of god, and for the testimony which they held.

10. And they cried with a loud voice, saying how long o lord, holy and true, dost thou not judge and avenge our blood on them that dwell on the earth?

11. And white robes were given unto every one of them, and it was said unto them, that they should rest yet for a little season, until their fellow servants also and their brethren, that should be killed as they were, should be fulfilled.

12. And I beheld when the lamb had opened the sixth seal, a great earthquake, and the sun became as black as sack cloth of hair, and the moon became as blood.

13. And the stars of heaven fell unto the earth, even as a fig tree cast down her untimely figs, when she is shaken of a mighty wind.

14. And the heaven departed as a scroll when it is rolled together, and every mountain and island were moved out of their places.

15. And the kings of the earth, and the great men, and the rich men, and the chief captains, and the mighty men, and every bondman, and every free man, hid themselves in the dens and in the rocks of the mountains.

16. And said to the mountains and rocks, fall on us, and hide us from the face of him that sitteth on the throne, and from the wrath of the lamb.

17. For the great day of his wrath has come, and who shall be able to stand?

MATHEW 5:13

You are the salt of the earth, but if the salt has lost its savour, wherewith shall it be salted? It is therefore good for nothing but to be cast out, and to be trodden under the feet of men.

MATHEW 5:14

Ye are the light of the world, a city that is set on a hill cannot be hid.

MATHEW 5:15

Neither do men light a candle and hide it under a bushel, but on a candlestick and it giveth light to all that are in the house.

MATHEW 5:16

Let your light so shine before men, that they may see your good works, and glorify your father, which is in heaven.

STEVEN SCHOOL

The first book of Moses called
GENESIS.

1. In the beginning God created the heaven and the earth.

2. And the earth was without form, and void, and darkness was on the face of the deep. And the Spirit of God moved upon the face of the waters.

3. And God said, let there be light; and there was light.

4. And God saw the light, that it was good; and God divided the light from the darkness.

5. And God called the light day, and the darkness he called night. And the evening and the morning were the first day.

6. And God said, Let there be a firmament in the midst of the waters, and let it divide the waters from the waters.

7. And God made the firmament, and divided the waters which were under the firmament from the waters which were above the firmament; and it was so.

8. And God called the firmament heaven, And the evening and the morning were the second day.

9. And God said, let the waters under the heaven be gathered together unto one place, and let the dry land appear, and it was so.

10. And God called the dry land earth, and the gathering together of the waters he called seas, and God saw that it was good.

11. And God said, let the earth bring forth grass, the herb yielding seed, and the fruit tree yielding fruit after his kind, whose seed is in itself, upon the earth, and it was so.

12. And the earth brought forth grass, and herb yielding seed after his kind, and the tree yielding fruit, whose seed was in itself, after his kind, and God saw that it was good.

13. And the evening and the morning were the third day.

14. And God said, let there be lights in the firmament of the heaven to divide the day from the night, and let them be for signs, and for seasons, and for days, and years.

15. And let them be for lights in the firmament of the heaven to give light upon the earth, and it was so.

16. And God made two great lights, the greater light to rule the day, and the lesser light to rule the night, he made the stars also.

17. And God set them in the firmament of the heaven to give light upon the earth.

18. And to rule over the day and over the night, and to divide the light from the darkness, and God saw that it was good.

19. And the evening and the morning were the fourth day.

20. And God said, let the waters bring forth abundantly the moving creature that hath life, and fowl that may fly above the earth in the open firmament of heaven.

21. And God created great wales, and every living creature that moveth, which the waters brought forth abundantly, after their kind, and every winged fowl after his kind, and God saw that it was good.

22. And God blessed them, saying be fruitful and multiply, and fill the waters in the seas, and let fowl multiply in the earth.

23. And the evening and the morning were the fifth day.

24. And God said, let the earth bring forth the living creature after his kind, cattle, and creeping thing, and beast of the earth after his kind, and it was so.

25. And God made the beast of the earth after his kind, and cattle after their kind, and everything that creepeth upon the earth after his kind, and God saw that it was good.

26. And God said, let us make man in our image, after our likeness, and let him have dominion over the fish of the sea, and over the fowl of the air, and over the cattle, and over all of the earth, and over every creeping thing that creepeth upon the earth.

27. So God created man in his own image, in the image of God created he him, male and female created he them.

28. And God blessed them, and God said unto them, be fruitful and multiply, and replenish the earth, and subdue it, and have dominion over the fish of the sea, and over the fowl of the air, and over every living thing that moveth upon the earth.

29. And God said, behold, I have given you every herb bearing seed, which is upon the face of all the earth, and every tree, in the which is the fruit of a tree yielding seed, to you it shall be for meat.

30. And to every beast of the earth, and to every fowl of the air, and to everything that creepeth upon the earth, wherein there is life, I have given every green herb for meat, and it was so.

31. And God saw every thing that he had made, and behold, it was very good. And the evening and the morning were the sixth day.

THE FIRST SABBATH.

Thus the heavens and the earth were finished, and all the host of them.

2. And on the seventh day God ended his work which he had made, and he rested on the seventh day from all his work which he had made

3. And God blessed the seventh day, and sanctified it, because that in it he had rested from all his work which God created and made.

4. These are the generations of the heavens and of the earth when they were created, in the days that the lord God made the earth and the heavens.

5. And every plant of the field before it was in the earth, and every herb of the field before it grew, for the lord God had not caused it to rain upon the earth, and there was not a man to till the ground.

6. But there went up a mist from the earth, and watered the whole face of the ground.

7. And the lord God formed man of the dust of the ground, and breathed into his nostrils the breath of life, and man became a living soul.

8. And the lord God planted a garden eastward in Eden, and there he put the man whom he had formed.

9. And out of the ground made the lord God to grow every tree that is pleasant to the sight, and good for food, the tree of life also in the midst of the garden, and the tree of knowledge of good and evil.

The garden of Eden is guarded by a flaming sword which points in all directions to protect the tree of life, and the tree of knowledge.

King and Queen, Sun and Moon, the male, and the female. Philosophical gold, and living silver. The key to eternal life, and wisdom.

THE FOUNTAIN OF YOUTH.

The fountain of youth is said to be a legendary spring which contains a special kind of water which is not ordinary water, but the water of life. Michael Sendivogius called this substance water which does not wet the hands. Some have said that this water is that river which flows forth from the garden of Eden and that it was placed there by the hand of God. The ancient sages called it the first matter of all things, and they remained steadfast and firm in their belief that all living things contain this substance within themselves. It is the universal spirit, the winged serpent that moves within the earth, and brings forth life to all living things which were created by the hand of almighty God himself.

This secret fountain, is said to heal all sickness and disease, and to restore the youth of all who drink of it or bathe in its divine waters.

The ancient sages believed that one must cross the land of darkness to find this restorative spring which is known as a universal panacea, a universal medicine, Aurum Solis, which means literally, Gold Of The Sun.

This elixir of life which is referred to in Genesis as the mist rising up out of the earth has often been equated with the philosophers stone.

The land of darkness that one must cross over in this journey has also been called the black earth of the Egyptians, as well as the black dragon. It has been said in ancient writings that one must slay this dragon with the white hot spear, which is the element of fire. Medieval alchemists believed in the four elements, Earth, Water, Fire, and Air. One element acting upon another, begins the creation of all things. Air, is the breath of God. Fire, acting upon Earth, produces Water, for it causes the mist to rise from the earth, where it mingles with the air, and ignites the divine spark of creation.

The first governor of Puerto Rico, Juan Ponce De Le'on, set sail many moons ago, in search of this mysterious water, the fountain of youth. In 1513 he landed at St Augustine Florida and traversed the creeks and streams of that area seeking the legendary water of life. The city of St Augustine is now home to the fountain of youth archaeological park which was built in 1904. Over the years remnants of early Christian life have been unearthed in the vicinity of this place. Many persons believe that this Spanish conquistador, Ponce De Le'on, never found the fabled fountain of youth.

The Wind Hath Carried It In Its Belly.
The Earth Is Its Nurse.
Its Strength It Doth Aquire In The Fire.
And There,
Becomes The True Stone Of The Ancient Sages.
Born as it were,
Of Fire, Earth, Water, and Air.
The Stone Of Fire.
The Secret Water Stone of the Wise.
The Stone Which Bears The Sign of the Sun.

The term Philosophers Stone means, the stone which bears the sign of the sun. It is the blessed corner stone, which the builders rejected. It has been written, that he who possesses this stone, can turn the water into wine, heal the sick, feed the hungry, restore youth, and overcome death itself.

Many times over the centuries people from all walks of life have pondered the question, why are there recordings of persons in biblical times living to around one thousand years of age?, why is it so different now?, did they record time differently back then?, is it smog?, pollution?, poor diet?, lack of exercise?, smoking?

Or was it,

Aurum Solis, Gold Of The Sun.

What would possess so many, ancient sages, from all walks of life, and from across all parts of the globe, to leave us so many treatises, pacts, and journals of their writings of ancient wisdom and philosophy upon this subject? It occurs to me that it would be one single thing, and of great importance.

VITRIOL.

The term Vitriol, does not refer to an actual substance, which many have believed.

Instead, this term is an abbreviation for a secret code which is written in Latin. This code reveals the process for the confection of the lapis philosophorum, which is the philosophers stone. Let us now, take a deeper look. Vitriol means, Visita Interiora Terrae Rectificando Invenies Occultum Lapidem. In English, the basic translation is this, visit the interior of the earth, and there, by rectification, find the occult lapis.

And now in simple terms, rectification simply means purification. In Alchemical work, we have three basic forms of this, and they are, Distillation, Extraction, and Calcination. Now during the course of your work, if your substance contains the vital life force energy or revivifying spirit, then we would not want to destroy this particular phase of our work with the dry heat of calcination, therefore we would choose either distillation, or extraction, depending on whether the matter to be purified is volatile, or fixed, will determine for us whether we will distill or extract our matter. Distilling involves the evaporation of a substance over the helm and into a receptacle, you could probably find more difficult terminology to phrase it with, but let us keep it simple. Extraction involves dissolving a substance into a liquid then filtering the liquid, and then evaporating it to dryness in order to recover the purified matter from the solution, just as if you were to evaporate sea water to obtain salt crystals, you could then redissolve them in distilled water, filter it, and repeat the evaporation process a few times to clean and purify the salt. You might calcine (bake) the dry salt to burn away any combustible impurities between the dissolutions and evaporations to achieve the optimum level of purity so that when elements are reunited they can become an exalted substance.

"VISIT THE INTERIOR OF THE EARTH"

It is not the earth itself that we are to be concerned with here, but something taken from it.

Here is a basic example of an Alchemic processing of a substance.

Imagine if you were to take sea water, distill it, capturing the distilled liquid in a suitable receptacle, then calcine (bake) the leftover matter which in our case would be sea salt, to remove its impurities, then add this calcined salt back to the same water which was distilled from it, now filter this solution, and then redistill it. Each time that you reiterate the same process you are further purifying and exalting both the salt, and the water. The ancient sages called each repetition of this mercury of one eagle, since the water itself is fluid like mercury, although it has nothing at all to do with common mercury, each repetition of this technique can also be referred to as one turn of the alchemical wheel. Each turn of this wheel brings our substance to a higher, and more exalted state of being, better than nature left it, climbing the philosophical ladder, higher and higher, reaching for the sun, the ultimate level of fixity and purity, which is found only in gold. The only element with a right hand atomic spin. Lead, is a soft metal, it has a left hand atomic spin, the atomic number of lead is 82, because it contains 82 protons. Gold on the other hand is atomic number 79, which means that it contains 79 protons. So these two metals are very similar in structure, lead had an additional 3 protons, protons can be removed by reversing the atomic spin of an element, reversing its polarity so to speak. If the left hand atomic spin of lead is reversed, the substance has no choice but to become gold. In the scientific world this generally requires a vast amount of energy, which renders the process to be a financial loss, not to mention the possibility of the small amount of resulting gold being radioactive and therefore worthless.

In the medieval realm of alchemy, the practitioners of the art did not have nuclear reactors or access to such vast amounts of heat and energy which modern technology has developed, instead their methods were simple, their subject of focus was very advanced for their time because the transmutation of one element into another is quantum physics, however their process would be more in line with what is now known as cold fusion, which requires much less heat and energy.

In alchemical writings, much terminology is used and during the course of this book will will in fact explain some of it. For instance, the term digestion, this simply means apply heat to the matter being worked upon, the term coction, means the same thing, example phrase, "subject the matter to coction", it simply means to cook your alchemical substance.

Wind furnace, now this may sound like a mythical object from the film industry, but instead it is very real and was of the utmost important to the medieval alchemists since they did not have modern gas fired high heat furnaces to work with, the wind furnace is simply a wood burning furnace, in which a bellows can be incorporated in order to increase the heat.

Athanor, this is simply a wood burning fireplace made of clay bricks, it is circular in shape, and constructed like a tower, it has a fire grate inside. The top of this tower is the chimney, this type of alchemical furnace is used for baking and roasting, which is to calcine the matter.

Putrefaction. This is known as the black stage of the great work, many persons have different ideas about this, so let us examine it a little closer. It is said that this portion of the great work occurs in the philosophical egg, and most people believe that means a round bottom flask, which is incorrect. Maria the prophetess used a distillation vessel called a Tetrabiko's, this is a medieval rendition of a modern column still. In common terminology there are basically two types of stills used for distilling of various substance, the first is the pot still, the alchemical version of this is called a retort, this will suffice to distill the matter, however a column type of still will bring the matter to a much higher level of purity and Mary's Tetrabiko's is the alchemical version of this type of distillation vessel. Having a long neck, versus a short neck, for the ascending vapors to climb before finding their way over the helm and into the receptacle, will exalt our alchemical substances to a much higher degree of purity. So the black stage of the great work of the magnum opus, is actually just a simple distillation, when you hear the term putrefaction, or the ravens head, simply envision the distillation of matter in the same way that alcohol is made. So the true philosophical egg, is the distillation vessel, the most common of these in alchemy is the retort. The round bottom flask is still needed but the alchemical name for the flask is actually the oven of secrets because that is where our substance is baked and cooked, or "digested" by heat. The matter does in fact turn black during this portion of the work as most substances which are being distilled will do. At the beginning of this book we noticed that in Genesis, the lord god talks to us about the mist ascending from the earth, which describes very well what we have discussed here about alchemy in regards to the actual performance of the magnum opus of the great work.

For those of you who are more experienced alchemists, at some point in your work if you are to proceed, you will find yourself engaged in transmutation experiments, you probably will not run down to your local pawn shop babbling to them that you are an alchemist and you want them to test your recently transmuted elements for you, so here I will list some simple, home testing methods for determining the success and quality of your work which will help you to proceed in the right direction. First of all, silver. When silver is banged against another metallic object it should have a ring to it, similar to the way that a bell works when you strike its surface with a piece of metal.

Ice works as well, but in a different way, simply set an ice cube on to the piece of metal that you believe might be silver, if the metal really is genuine silver, the ice cube will begin to melt immediately.

The Latin term argentum is commonly found in alchemical writings, it refers to silver, to which is assigned the atomic number of 47, which is related to the number of protons which it contains. It is assigned the symbol Ag. Out of all of the elements, silver has the highest electrical conductivity, gold on the other hand, (atomic number 79), is the only element with a right hand atomic spin. Silver is slightly harder than gold and its melting point is 961 Celsius, 1761 Fahrenheit, which is another way to test it by determining if it melts at the correct temperature.

You can also purchase a diamond tester kit online for about fifty dollars American currency, which can be used at home to test gold, silver, platinum, and diamonds, as well as a variety of other metals and stones.

The melting point of gold is 1063 degrees Celsius, or 1945 degrees Fahrenheit.

In the realm of mineral alchemy, gold represents the ultimate degree of purity and fixity.

SILVER AS MEDICINE.

Silver has been said to possess antibiotic as well as antimicrobial qualities, it was used during the civil war for its medicinal properties. Earl settlers travelling across America used to keep a silver coin in their milk, which was said to keep it fresh longer. During civil wars times many soldiers kept a silver coin in their water canteens. Excessive consumption of silver can cause a blue tint to the skin which is where the term "the blue bloods", came from.

I myself tried colloidal silver and found that it improved my night vision. In alchemy, the modern symbol for the plus sign has been associated with silver for many centuries, this is an ancient symbol for salt, also the color blue is used which signifies water.

The moon crescent symbol is also associated with silver, as well as the terms Luna, Beya, Diana, Doves of Diana, Diana unveiled, living silver, philosophical silver, aqua vitae, and the water of life.

All of these terms and symbols have been equated with the elixir of life, as well as symbols for the sun, and the metals gold and silver.

Many of the ancient sages suggested in their writings that they believed there was a great medicine for man hidden in the oil of sulfur, or a derivative of this oil. the work of obtaining the oil of sulfur in the traditional methods is outlined in some alchemical manuscripts from previous centuries, great caution is advised here, because the proper method can be hazardous if improperly done, as with many other laboratory procedures.   As with many hermetic operations, there are two methods to obtain the oil, the first is distillation and produces the purest substance, however this is also the most dangerous, it is referred to in hermetic science as a "dry path", the second method is called extraction, this takes longer, and the substance is not as pure, but it is a safer technique to perform, and this method is called a "wet path, wet method, or wet way", in laboratory alchemy.

Centuries ago, alchemists were believed to have collected morning dew in spring time, as well as rain water, especially rain water collected during a thunder storm. For years, many persons have speculated about this, is this rain and dew the substances which were used to create medicines for man and metals?, are these the ingredients of the philosophers stone or the elixir of life?, in short, No.

Many scientific experiments require the use of distilled water for its purity, especially freshly distilled water since it is not only pure, but also highly oxygenated. This explains why they were adamant about collecting the water before it touched the ground, it was also used to wash and clean alchemical vessels and apparatuses since it tends to be free from many of the minerals and impurities which are commonly found in ground water.

Rain water being clean water, and oxygenated, would have some health benefits as long as it is not polluted, as in the case of acid rain.

Oxygen, symbol O, atomic number 8, is the most abundant element by mass within the earth's crust. It is a gaseous vapor which can diffuse through plastic bottles over time which means that bottled water which is stored in plastic will lose its oxygen content over time.

If these plastic bottles are left in the sun the plastic begins to break down and the water will absorb some of the chemical properties from the decomposing plastic, which is not healthy to drink.

A glass of water set on a table or counter top will absorb oxygen directly from the atmosphere, many alchemists believed that this phenomena increases the life force energy within the water.

The human body is composed mainly of water, the interesting thing here is that as we age our water level decreases, and the older we get, the less water our bodies contain, it would seem to me that loss of water in the human form has something to do with the aging process. Salt causes our bodies to retain water, in Mathew 5:13 Jesus has told us,

Ye are the salt of the earth; but if the salt have lost his savour, wherewith shall it be salted? It is thence forth good for nothing, but to be cast out, and to be trodden under the foot of men.

Medieval alchemists believed that excess affected the longevity of one's life span, whether it be excess of food, excess of alcohol, or excess of tobacco. This theory also included excess of heat, and excess of cold in the items being consumed, as well as excess of spicy things like hot peppers.

They believed that consuming foods which were very hot as well as beverages that were excessively cold would reduce the life span of the individual.

They were striving for a happy medium, choosing foods that were not too spicy, not too hot or cold, and consuming them in smaller quantities as well as making sure to get a wide variety of foods in their diet. For example, if they were to eat a slice of tomato which is acidic, then directly after they would consume something non acidic to counteract the acidic effects of the tomato upon the human body, and vice versa.

It would seem to me that they were trying to maintain a neutral or balanced Ph level within the water contained in their bodies.

A fellow alchemist recently asked me, after death of the human body, which would you prefer, cremation with ashes scattered to the four winds, cremation with ashes kept in a container, burial with embalming, or burial without embalming?

To this I reply,

Let the fire reduce us back to our first matter, let the wind carry us in its belly and return us to the earth from whence we came. Let the water fall from the heavens and bring the revivifying spirit back to us.

In Jesus name we pray, Amen.

STEVEN SCHOOL

VERNAL EQUINOX

The vernal equinox occurs twice per year, it is a Latin term which means equal night, because around the time when the equinox occurs both day and night are about equal length which is caused by the earths equator passing the center of the sun.

The great sphinx on the Giza plateau in Egypt points directly towards the rising sun on the day of the vernal equinox and this marks the first day of spring which occurs around march 20, and the first day of winter which occurs around September 22.

To the alchemists of medieval times, these days had significant meaning, they had a secret recipe for a long and healthy life, they would take one grain of the philosophers stone dissolved in white wine, the medicine was diluted with this wine (which had to be made from grapes) until it was the color of liquid gold and contained no redness from the celestial ruby. This liquid was allowed to stand until a whitish ring appeared around the edges of the glass, at which time the liquid was filtered to remove the white substance.

This medicine was taken on the vernal equinox twice per year in the dose of two drops, and this was held to be a great secret which they called the token of truth.

The interesting thing about the fact that the wine had to be made from grapes as opposed to other distilled spirits is that grapes contain antioxidants which slow down or retard the aging process, some grapes (mainly the darker colored varieties such as red or purple), also contain a substance called resveratrol which is good for skin and is known to be an anti inflammatory.

THE ALKAHEST.

The alchemists had a term which was known as the alkahest, this substance was said to have the ability to dissolve metals and even gem stones back to their first matter, which is what they were in the beginning, before they formed, they believed that this could lead to a process of multiplying and exalting these substance, or even to turn ordinary minerals into diamonds and the like. Over the centuries many persons have speculated about what this alkahest substance could be, many have believed that it is some type of crystalline salt.

It has been well known that these ancient scientists experimented with many substances including urine, which has caused many modern aspiring alchemists to believe that the philosophers stone was actually made from urine which has sufficiently been proven to be false.

The truth of it is that the alkahest is in fact a crystalline salt, it is not activated until it is added to a certain water (not ordinary water) which is known to contain life force energy, some have equated this water to be that spoken of by Michael Sendivogius which he called "water which does not wet the hands", this assumption is also false, Michaels water which does not wet the hands is an entirely different substance, and for a completely different purpose.

The alkahest is directly related to the urine work which was done by the medieval alchemists, they would place urine in glass jars and allow it to putrefy or ferment, this substance was very foul smelling, then they would distill this matter with very low heat, like that of a hatching chicken, using a glass retort. The clear distillate which came over the helm into the receptacle (the most volatile portion), was saved in a sealed glass flask, the remaining substance in the retort was poured out and evaporated to dryness. It was then distilled in an aludel to collect and purify the volatile salt of urine, this pure crystalline salt was the alkahest, when it was added back to its own clear distillate and digested, it became very penetrating and was even said to dissolve the glass if it wasn't given something to act upon such as metals or gem stones.

The alchemists meaning of digestion was to apply heat to the substance, usually in a sand or water bath, and this was a gentle cooking, not burning the matter, the water bath was developed because it does not allow the heat to rise above the boiling/vaporization temperature of the water. The sand bath was generally the temperature of warm sand, like sand in sunlight on a summer day.

The hand of the philosophers.

THE PARTING OF THE RED SEA.
EXODUS 14:21

And Moses stretched out his hand over the sea; and the lord caused the sea to go back by a strong east wind all that night, and made the sea dry land, and the waters were divided.

EXODUS 32:20
THE GOLDEN CALF.

And Moses took the golden calf which the people had made, and burnt it in the fire, and ground it to powder, and strewed it upon the water, and made the children of Israel drink of it.

What is the "fire", but the water?

What is the "gold" but the salt?

STEVEN SCHOOL

The parting of the red sea.

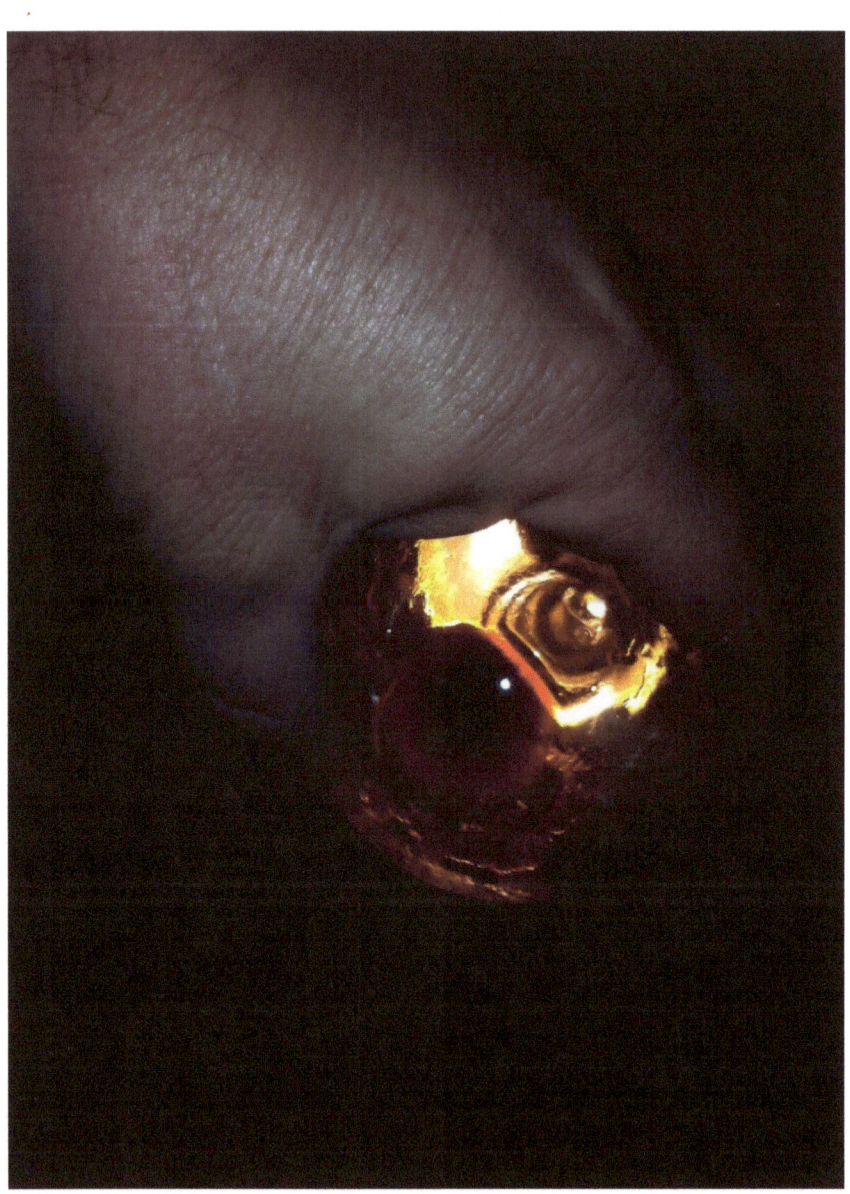

At the beginning of this book, in "the seals" number 2, behold a white horse, and a crown was given unto him. We can compare this to the hand of the philosophers which is in this book, we can see there that the thumb of the hand has the crown and the moon symbol, it signifies this white horse.

In the chymical wedding of father Cristian rosencreutz, father CRC tells us, join the red man to the white wife.

In the "seals" number 4, and there went out another horse which was red, and that they should kill one another, and there was given unto him a great sword. This refers to the chymical wedding of Cristian rosencreutz, which the ancient philosophers called conjunctio, which simply means conjunction. It is the combining of purified substances, to form an exalted substance, or stone, such as the philosophers stone. The matters to be combined, began as one substance in the beginning, the original matter or prima materia, was put through the alchemical process to reduce it back to its first matter, separate and purify the elements, then recombine them, nothing enters into the composition, which did not originate from it, in the words of Michael sendivogius which he has left us in his new chemical light,

"If anyone, for instance, were to attempt the creation of a man out of a man's hand and a woman's foot, he would fail."

He is telling us here, the same thing, in the conjunction phase of preparing alchemically exalted matter, nothing foreign must enter into the composition, or the work will become a failure.

In "the seals" number 12, "and the sun became black, and the moon became as blood", this statement identifies for us two substances, and tells us which one is the sun, and which one is the moon, the statement also matches what occurs during the distillation of matter, the volatile portion or liquid evaporates over the helm and collects in the receptacle, in alchemy the moon is associated with water, blood and water are both liquid, the distilled liquid is the "blood", of the substance that it was distilled from, the capuut mortuum, which is the matter left in the distillation vessel becomes a charred black matter, once the liquid has left it, the heat used for the distillation causes the subjects combustible sulfur to become burned, this is the dark land that we had to cross over to find the waters of life, this is the "black earth of the Egyptians" which we have spoken of previously in this book.

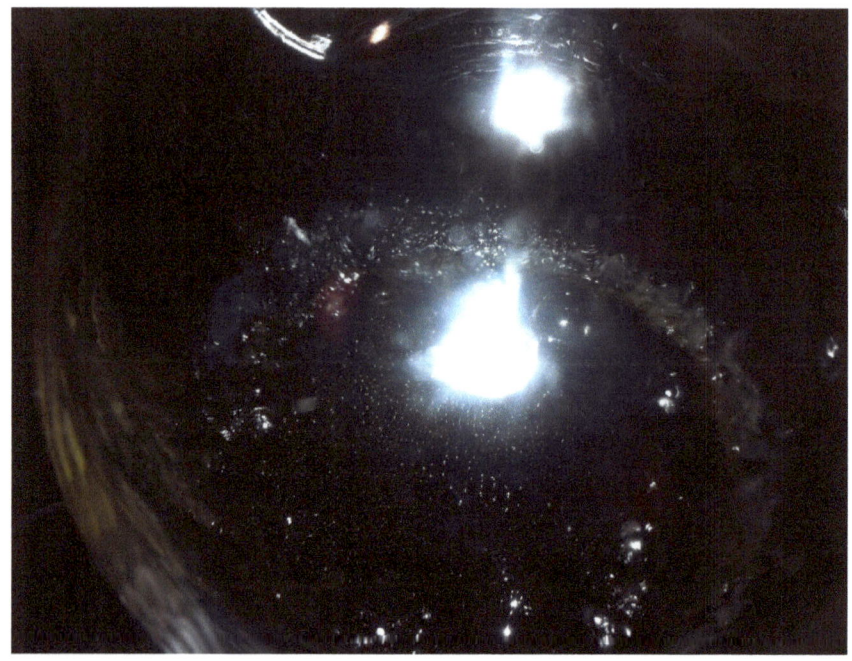

The Black stage of the great work, occurring in the philosophical egg.
"Separate the pure, from the impure, and with great industry".

Aurum Solis

Mars is the stout and valiant god of war,
His body vile, and little is esteemed,
He's fierce of courage, conquering near & far
all sturdy opposites, and may be deemed,
that his rough outside hidden doth enclose,
a spirit whose full virtue no man knows
Venus, the planet fair, the god of love,
Whose beauty the stout god of war allures,
her central salt; whoso has wit to prove,
shall find a key all secrets which assures
the owner for to find; I say no more,
for this disclosed by none, hath been before.
To Saturn, Mars with bonds of love is tied,
who is by him devoured of mighty force,
whose spirit Saturn's body doth divide,
and both combining yield a secret source,
from whence doth flow a water wondrous bright,

In which the sun doth set and lose its light.

In Genesis we see that first God created heaven and the earth, then God divided the light from the darkness. This means,

1. black, earth, darkness, sun, sol, gabritius, philosophical gold, the red horse, the male element.

2. light, heaven, water, blood, moon, beya, philosophical silver, the white horse, the female element.

In the words of Michael sendivogius, burn the earth until it becomes sulfur incombustible, for there is no danger here. When the black matter is calcined in a great fire it becomes an incombustible red powder, the "earth" contains an impurity which can only be removed by fire, and that of the fourth degree of heat, which means bathed in flame, this is the fire lizard of alchemy.

Join the red man to the white wife.

The red and white powders.

ATHANOR.

SULFUR INCOMBUSTIBLE.

The Waters of Creation.
Victory over Chaos.

The Celestial Ruby of Alchemy.

LIFE FORCE ENERGY.

All living things, be they plants, trees, animals, minerals, or humans, are animated by the same life force energy. Water is a magnet to this universal spirit. The energy level of the water is also affected by temperature. Cold water will contain more life force energy than hot water, as water becomes colder, its level of life force energy will increase, heat it up again and the energy level will decrease.

Set a glass of water outside on a table on a clear night, it will attract the universal spirit directly from the atmosphere and it will store it in the same way that a battery stores an electrical current.

Some liquids contain more life force energy than water, one such example is fresh cow's milk, freshly harvested and chilled. This milk is packed full of the universal life force energy. Look at the milk in your local grocery store, the milk has been processed, effectively removing most of the universal spirit, what is left can escape directly through the plastic container. It still does contain some of the life force energy, but nowhere even close to what it had when it was first collected.

Now let us look at the eggs, when they were freshly laid, they were absolutely brimming with the universal spirit of life, but after they have become three days old, the energy is gone.

The same is true with vegetables, if it is eaten directly after being plucked from the vine, it will contain the revivifying spirit to recharge and revitalize your body, but the vegetables in the store are completely dead three days after being harvested.

The true purpose of living beings consuming food and drink is to replenish our supply of the animating spirit which gives us life. The same is true in the case of minerals, water and oxygen is their nutriment which causes them to grow within the earth.

> God has told us in Genesis, let all things multiply in like kind,
> which applies to all things that God has created.
> Water and oxygen, the body of man, and the breath of God.

GENESIS 27:28

Therefore God give thee of the dew of heaven, and of the fatness of the earth, and plenty of corn and wine.

The corn and the earth are the same thing, which is not corn or earth, likewise the wine and the dew are but one thing, which is not wine,

Sun and Moon, Gabritius and Beya. The red man and the white wife.

Now let us look at the metals, before they are plucked out of the ground they are called ore. This is actually a living growing crystal, it contains the same universal life force energy as all living things. The water and minerals present in the ground are the nutriments of this ore, they cause it to grow just like stalactites and stalagmites in underground caverns.

If you pluck the metallic ore out of the ground, it will stop growing because you have removed it from its food supply, however this metallic crystal is still alive, if you place it right back into the ground in the same exact place that you plucked it from, it will resume growing. If you decide to melt this ore in a furnace, the heat of smelting will destroy the life of this metallic crystal and render it dead, as the temperature of the furnace rises, it expels the water which is trapped inside the ore, the very liquid which is infused with the life force energy, which is the animating spirit, it is the universal spirit of God which brings life to all things terrestrial.

Metals are alive when they are in their ore, the fire of smelting is their death. All living things are destroyed by excessive heat, even a tree, the reason being that first the moisture which contains the animating spirit is driven out, such moisture that protects the subject from all heat and excess has left the body, then the matter itself which is the substance of the body is damaged by the excessive heat, which means that the animating spirit cannot return and resurrect if the body has been destroyed, such as in the melting of an ore, or the burning of a tree.

TREATISE ON METALS.

Metals are composed of minerals, impurities, and water. The basis of each metal is its fixed salt, this substance the alchemists called sulfur.

Of this fixed salt there are two types, the white and the red. The white salt is the basis of the white metals, the red salt is the basis of the red metals, water is the mineral seed which causes these living crystals to grow and multiply within the earth.

The metals also contain a volatile salt, the fixed salt of metal is incombustible and unaffected by fire, other than to say it will melt and become like glass. The volatile salt cannot be burned, in heat it simply volatilizes, and distills away, (sublimes) if it is collected in a receptacle, such as in a distillation per aludel, it adheres to the glass and coagulates back into volatile salt, and this method can be used to purify the substance.

If the fixed salt of metals is to be purified, it is calcined (baked) to remove impurities, which are called combustible sulfur, then the method of extraction is used to cleanse the matter, which means that it is dissolved in a suitable liquid, then filtered and evaporated back to dryness.

When nature creates a metal in the earth it is called a metallic ore which is a living and growing crystal, which metal is formed depends upon which fixed salt is present within the earth's crust, as well as whether or not there are any impurities present.

If the ground is pure and the white salt is present, silver will be produced, if the red salt is present and the earth is pure then gold will be created.

When impurities are present a lesser metal will grow such as lead, which has the atomic number of 82, lead also has a left hand atomic spin.

Gold is atomic number 79, and it has a right hand atomic spin. The elements are assigned an atomic number based upon how many protons they possess. Lead has three more protons than gold, if you were to reverse the atomic spin of lead, it would have no choice but to become gold.

In nuclear physics, a vast input of energy is required to reverse the atomic spin of lead and to cause it to lose three protons. In the realm of alchemy however, which is an ancient and primitive art, cold fusion is the goal. Two methods, one nuclear and one natural.

Many persons are familiar with a report that Russian scientists working on an experimental nuclear reactor near lake Baikal accidentally discovered a process for turning lead into gold when the lead outer casing of their nuclear reactor suddenly changed into gold after the device had been activated in 1972.

In modern science, elements can be transmuted with the use of particle accelerators which is quite common although it is quite costly since the expense of it outweighs the potential profit.

Over the centuries many alchemists have been asked, what are the ingredients of the philosophers stone, to which some of them have replied minium and litharge which are code words designed to mislead the multitude.

In the great work of alchemy we often see pictures or written references to two lions, one green and the other red.

The term litharge refers to the green lion, just as the term minium refers to the red lion these are the gateways in the alchemical path. This may greatly simplify things for you if you are reading ancient alchemical texts which use this type of terminology.

Sun and Moon, King and Queen, dancing in the fire, sunlight mingled with the light of the moon.

STEVEN SCHOOL

GROWING CRYSTALS.

The philosophers stone has been called Sal petrae, which means salt stone. It would seem prudent to assume that it would therefore be an alchemically exalted stone made from some type of salt. Most people would agree that to repair a rubber tire which has gone flat, you generally might require a rubber tire patch, as well as rubber cement to glue and bind the rubber patch to the rubber tire, therefore we can assume that a remedy for the metals would come from the metals themselves.

In the writings of Michael sendivogius he has told us that the life of the metals is in their ore, and that the fire of smelting is their death. See below,

"If any one, for instance, were to attempt the creation of a man out of a man's hand and a woman's foot, he would fail. For there is in every body a central atom, or vital point of the seed, its 1/8200 part, even in a grain of wheat. Neither the body nor the grain is all seed, but in every body there is a small seminal spark, which the other parts protect from all excess of heat and cold. If you have ears and eyes treasure up this fact, and be on your guard against those who would use the whole grain as seed, and those who strive to produce a highly rarefied metallic substance by the vain solution and mixture of different metals. For even the purest metals contain a certain element of impurity, while in the inferior the proportion is greater. You will have all you want if you find the point of Nature, which you must not, however, look for in the vulgar metals; it is not to be found therein, for all these, and common gold more especially, are dead. But the metals which we advise you to take are living and have vital spirits. Fire is the life of metals while they are still in their ore, and the fire of smelting is their death."

He is here illuminating the path that one would take in the great work. The alchemists believed that the metals (while still in their ore), were composed of salt, sulfur and mercury, being that these are medieval alchemical terms, philosophically describing the substances they were working with, these terms are not to be confused with common salt, sulfur or mercury.

Modern science has developed simple as well as advanced methods for growing crystals at home or in the laboratory. Some of the more advanced methods might be used to create emeralds, rubies, or diamonds. Rubies are formed from alumina, diamonds are formed from carbon. These types of crystals are usually formed under intense heat and pressure.

Here are two basic methods of growing crystals at home, the seed crystal method, and the substrate evaporation method.

In the first method a seed crystal is required, it is created from a saline aqueous solution, a few drops of the liquid are placed onto a suitable evaporation dish such as a plate. This small amount of the saline solution is then very gently evaporated to dryness at low temperature, like that of warm sunlight. Once this liquid has evaporated, the dry matter which is left on the evaporation dish is the seed crystal. These seed crystals are then added back to the aqueous solution from which they came, causing it to become super saturated with its own salt, a glass jar is suitable for this portion of this type of experiment. The top of the jar is covered with a paper towel or a coffee filter to keep the dust out as these salt crystals grow and multiply, with the top of the vessel open in this fashion, the substance has exposure to oxygen and the excess liquid can evaporate to dryness as the crystals are formed. Once the experiment is complete, more of the same aqueous solution can be added to the jar in order to "multiply" these crystals. This method is very simply, to some people, alchemy may seem very advanced, in my opinion the medieval alchemists were simple and primitive persons working with rudimentary techniques, in the alchemical writings of Mr Sendivogius we find this,

"Simplicity is the Seal of Truth"

In the substrate/evaporation method of growing crystals a substrate is used such as charcoal for example. The charcoal can be broken up into chunks which is then placed into a shallow glass evaporation dish such as a glass pie pan. Some of the saline solution is then poured over the charcoal, as the liquid slowly evaporates at room temperature, the crystals begin to form and grow on the substrate.

The mineral corundum is fairly common in the earth's crust, it can also be found in many types of rock. In rare circumstances if the conditions are perfect during the formation of these crystals, a transparent or translucent gem stone may be produced by nature. If the stone turns out clear, it is simply corundum, if it is of gem quality and colored then it will either be a ruby or a sapphire. The dark red variations being the rubies and all other colors belonging to the sapphire family. It has been said that the finest quality rubies generally come from Burma. The color of sapphires usually ranges from blue, yellow, and orange, to green.

Corundum itself is made from aluminum and oxygen. It is written scientifically as A1203. If the resulting crystals produced are translucent or clear, then they are of gem stone quality, if they are opaque then they may be considered as industrial quality. As these crystals are forming in the earth which finished crystal is formed depends upon which elements are present, some elements have a coloring effect, the opaqueness of the lesser quality opals and rubies is caused by sodium being present in the equation. If there is salt present during the formation of this matter, it acts as an impurity and causes the opaqueness in the finished crystal.

If finely ground alumina is melted it reforms into a large crystal upon cooling, the temperature required to melt this substance is 2000 degrees Celsius, or 3600 degrees Fahrenheit. If the finely ground alumina is free of sodium impurities the resulting crystal will be translucent.

Let us consider the elements regarding the creation of matter. If we were to bake a cake, we might first collect purified and prepared elements, such as flour, yeast, sugar, and water. Thus having possession of these four elements, we can combine them in different amounts and subject them to cooking, all we must do now is to sit back and wait for the harvest, but if the amounts of ingredients which were chosen is wrong, we may produce bread, muffins, or cookies instead of our desired cake. If the degree of heat used is incorrect, then either our finished substance will not form, or it will be destroyed by excessive heat.

Thus we can produce many things in alchemy of the four elements, the theory is the same. The metals are living growing crystals. They contain the very same animating spirit, (life force energy) as you and me. The living and vital spirit which animates all things, which however should not be destroyed with excessive heat.

So a thing begins, and so it ends.

The Lord's Prayer.

Our father
Who art in heaven
Hallowed be thy name
Thy kingdom come
Thy will be done
On earth
As it is in heaven
Give us this day
Our daily bread
And forgive us our trespasses
As we forgive those who trespass against us
And lead us not into temptation
But deliver us from evil.
AMEN.

OTHER BOOKS I HAVE WRITTEN.

Alchemy and the green lion, the truth of the philosophers stone.
Alchemy and the golden water, Illuminating the celestial ruby
Alchemy survival guide, truth illuminated
Alchemy and the peacocks tail, the degrees of the fire
Alchemy and the ravens head, the secret of the red mercury
Alchemy and the golden process
Alchemy and the tincture of gold
Sol and Luna, the hermetical wedding

The Magnum Opus DVD. www.createspace.com/381734
Alchemy and the athanor DVD journey into the fire
www.createspace.com/388897

The secret recipe book, kitchen tool box
Casino survival guide, breaking the bank
Chinese takeout recipes
How to make money
Karate secrets revealed, Knowledge of the masters
The Kitchen Ninja
The Kitchen Ninja 2, Mexican Cuisine
Grandmas Delicious Recipes
Trophy Wife
Booze Survival Guide
Wilderness survival tips
Kitchen survival guide
Tools of The Trade, secrets of book promotion
How to promote a book, Winners Circle

ABOUT THE AUTHOR

I enjoy cooking at home from scratch, I like writing and have been interested in becoming an author for many years, thanks to amazon my dream has become a reality. I enjoy the company of my Rottweilers, I also enjoy the hermetic arts and have been studying alchemy diligently since 2008 with an emphasis on the white and red philosophers stones, the elixir of life, and the Primum ENS Melissa.

www.ingramcontent.com/pod-product-compliance
Lightning Source LLC
Chambersburg PA
CBHW040922180526
45159CB00002BA/571